Urgessa Tilahun

Review of Production, Productivity and Marketability of Soya Bean in Ethiopia

GRIN Publishing

Bibliographic information published by the German National Library:

The German National Library lists this publication in the National Bibliography; detailed bibliographic data are available on the Internet at http://dnb.dnb.de .

Imprint:

Copyright © 2014 GRIN Verlag GmbH
Print and binding: Books on Demand GmbH, Norderstedt Germany
ISBN: 978-3-656-89810-8

This book at GRIN:

http://www.grin.com/en/e-book/292692/review-of-production-productivity-and-marketability-of-soya-bean-in-ethiopia

GRIN - Your knowledge has value

Since its foundation in 1998, GRIN has specialized in publishing academic texts by students, college teachers and other academics as e-book and printed book. The website www.grin.com is an ideal platform for presenting term papers, final papers, scientific essays, dissertations and specialist books.

Visit us on the internet:

http://www.grin.com/

http://www.facebook.com/grincom

http://www.twitter.com/grin_com

Review of Production, Productivity and Marketability of Soya Bean in Ethiopia

Urgessa Tilahun Bekabil

Oromia Agricultural Research Institute, Haro Sabu Agricultural Research Center, Kellem Wollega, Dale Sadi District, P. O. Box 10 Haro Sabu, Ethiopia

Abstract

This review of research presents recent Soybean related studies conducted in Ethiopia. After a brief contextualization of the discourse regarding Soybean market chain, production and productivity researches globally, materials specific to Ethiopia is discussed in argument, synthesizing the types of findings, summarizing the trends and highlighting knowledge gaps. A review of this nature makes diverse research results available and accessible, facilitates knowledge translation and enables researchers to identify areas for future research.

Key words: Production, Productivity, Marketability, Soybean, Ethiopia

Introduction

The origin and early history of soybeans are unknown. It is not uncommon to read in agronomic publications that the earliest recorded origins of soybeans date back to 2800 B.C. in China [1]. Soybean, a short-day plant, is a very important oil and protein crop. It can grow on all types of soil, but deep fertile loam with good drainage is most suitable for growth [1]. The soybean (*Glycine max*) is one of the most important food plants of the world, and seems to be growing in importance. It is an annual crop, fairly easy to grow, that produces more protein and oil per unit of land than almost any other crop. It is a versatile food plant that, used in its various forms, is capable of supplying most nutrients. It can substitute for meat and to some extent for milk. It is a crop capable of reducing protein malnutrition. In addition, soybeans are a source of high value animal feed [2].

1

Soybean is an alternative protein source to the rural families and can be utilized at home in various forms and the surplus can be sold to other consumers and manufacturers for income [3]. Soybean is among the major industrial and food crops grown in every continent [4]. Soybean has an average protein content of 40% [5] and is more protein-rich than any of the common vegetable or animal food sources. Soybean seeds also contain about 20% oil on a dry matter basis, and this is 85% [4] unsaturated and cholesterol-free [3].

Soybeans have been variously referred to as the miracle golden bean, the golden nugget, nuggets of nutrition, pearls of the Orient, the cot; of China, the meat of the fields, the meat that grows on vines, the Cinderella crop of the century, the Cinderella crop of the West, the protein hope of the future, and the amazing soybean. Regardless of what they are called, soybeans are a promising and proven source of plant protein and edible oil [1]. Soybean (*Glycine max*), one of the world's major crops, has been cultivated by men since nearly 5000 years because of its agronomic and nutritional value. Soybean is a source of edible oil (second most consumed oil in the world after palm oil) and is used to produce livestock feed. Many other products with a soya basis are also directly used for human consumption (soymilk, soy yogurt, snacks, soya sauce, protein extract and concentrates, etc.) [5].

In the major producing countries and particularly in Brazil, Argentina, Paraguay and the USA soybean contributes significantly to the total value added by the agricultural sector. In these countries, soybeans and its sub-products also occupy an important position in total export earnings. Among smaller producers only India and Bolivia earn significant income from the exportation of soybean and derived products [6].

Review of Recent Research

Marketing Condition and Value Chain of Soybean in Ethiopia

Soybeans were tentatively tried in Ethiopia in the 1950s. A growers' manual was even published in Amharic and instructions on how to use the "foreign pea," as the soybeans were called at that time in Ethiopia, were also included [1].   Ethiopia with almost 80 million inhabitants is a large market for edible oil. The combination of a growing GDP with an annually population growth of 2.6% makes it even more relevant that the availability of good quality oil, such as soybean and sunflower oil, is improved. There is also a large scarcity in high protein animal feed for the booming dairy, export beef and poultry sectors. Similarly, there is strong demand from the nutritious food industries; factories that supply to the World Food Program alone have a total annual demand of 60,000 tons to cater for soy blends for the food insecure and malnutrition affected areas [7].

Access to local markets appears to be the main constraint in many developing countries in the tropics and sub-tropics where local soybean production could improve farmer incomes and the sustainability of the production system. Often soybean is imported into countries by the local vegetable oil and feed industries and as a consequence no demand for the crop is felt in the farming community. Where good market links from processors to local farmers have been made, as in Nigeria and especially in India, the farmers generally respond and the crop finds a good home in diverse cereal and root crop based production systems. Farmer incomes improve and the production systems become more sustainable. The rate of smallholder-based soybean production increase in India is one of the most remarkable stories in recent agricultural history. Many farm communities where the crop has found a niche have had substantive improvements in income and quality of life. Soybean can be a valuable alternative crop for many small-holder producers [6].

Production and Productivity Gaps of Soybean in Ethiopia

There are favorable climatic and soil conditions for soybean production in South and Western Ethiopia which is essential both for commercial purposes as well as for subsistence farming [7]. The problems of producing soybean is not only limited to market access but also low productivity and production, lack of processing facilities, lack of capital to increase production and no market information system for effective agricultural marketing [8].

Soybean is a high value and profitable crop. The economic viability of soy production is determined by the commercial utilization of both its sub-products, meal and oil, which, respectively, account for about two thirds and one third of the crop's economic value. Soy oil and meal are consumed worldwide as food and animal feedstuff respectively [6].

Soy meal accounts for over 60 % of world meal production (vegetable and animal meal) and occupies a prominent position among protein feedstuffs used for the production of feed concentrates. Soybean oil is the second most important vegetable oil (after palm oil); it accounts for 25 % of global vegetable/animal oils and fats consumption. The widespread use of soybean oil in particular as edible oil is due to its plentiful and dependable supplies, its competitive price,

3

and its neutral flavor and stability in both un hydrogenated and partially hydrogenated form. Moreover, the rapid rise in the demand for compound feed - und thus soya meal - has contributed considerably to the rise in soy oil production. Oil palm is a major competitor with soybean oil. Although the palm produces far more oil per unit area than soybean, soybean's role is expected to be secure because soya meal is in huge demand and oil is a very lucrative by-product. It is also true that oil palm is generally grown in different ecologies than soybean, so there is a certain amount of geographical complementarities [6].

Soybean varieties selected for drought tolerance have the potential of improving agricultural productivity and hence livelihoods if adopted by farmers [9]. Soybean grows in areas where maize and common beans are grown. It grows to a height of 60–120 cm, maturing in 3 to 6 months depending on variety, climate, and location. Soybean is drought tolerant. Depending on the variety, the crop can be grown from 0-2200m altitude and under rainfall ranging from 300 to 1200mm. Altitude influences temperature that in turn affects the initiation of flowering and maturity. At very high altitudes, flowering may not occur and the crop remains vegetative. Soybean is therefore a crop that requires warm climates and is suitable for low to medium altitudes [10]. It grows best when planted in pure stands. The presence of Rhizobium japonicum in the roots of soybean enables the crop to fix nitrogen in the soil contributing to improved soil fertility [11].

Table 1 Area under cultivation, yield and production of Soya Bean for 2005/06- 2007/08 main season Ethiopia

Year	Area (Hectare)	Yield (Quintals/Hectare)	Production (Quintal)
2003/04	1,027.00	4.46	4,574.00
2004/05	2,606.00	3.20	8,335.00
2005/06	3,327.00	11.46	38,119.00
2006/07	6,352.46	9.21	58,489.47
2007/08	7,807.40	10.80	84,006.39
2008/09	6,236.04	12.67	78.988.92
2010/11	11,261.12	14.05	158,244.22
2011/12	19,397.16	18.50	358,802.94
2012/13	31,854.75	19.98	636,531.01
2013/14	30,517.38	20.00	610,249.16

Source: Central Statistical Agency of Ethiopia

In spite of the importance of the crop and efforts made to enhance its production, the productivity of soybean on farmer's field has been low i.e., 3.20 – 20.00 quintals/hectare. According to the above table, table 1, from 31,854.75 hectares of land 636,531.01 quintals of soybean was produced during 2012/13 production period. This was the greatest production registered from 2003/04 to 2013/14. The average production therefore, is 19.98 quintals per hectare. The yield was increased to 20 quintals per hectare in 2013/14 from 4.46 quintals per hectare in 2003/04 production season.

Figure 1 Soybean production, yield and areal coverage

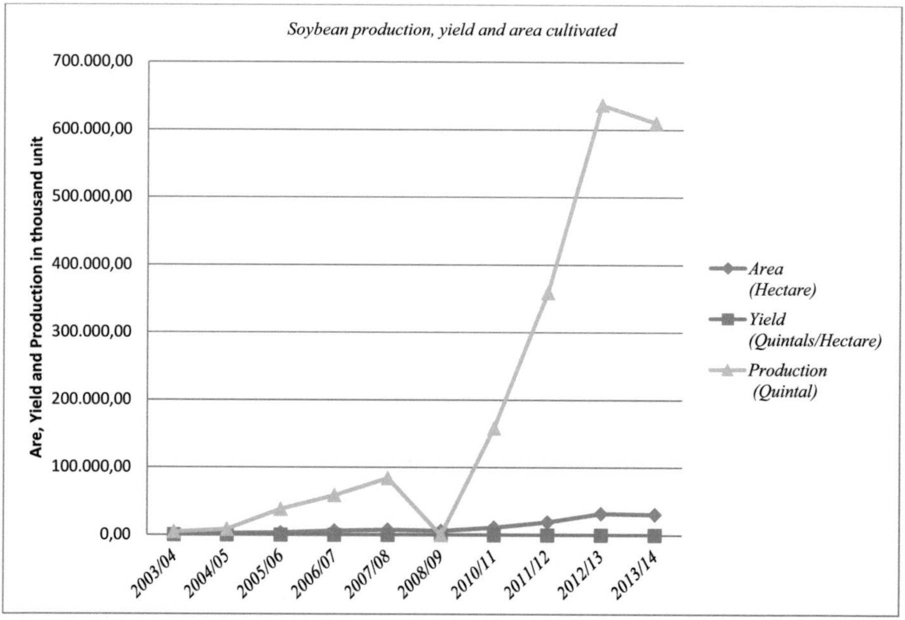

Source: Computed from table 1

The growth trend of production of Soya Bean in Ethiopia shows up and down movement during last ten years. This may be because of market problem of the crop, processing problem and unavailability of drought resistant and high yield crop varieties. The high production record was obtained in 2012/13 production season, which is 636,531.01 quintals. But high yield (20 quintals per hectare) was obtained during 2013/14 production season.

Table 2 Estimates of area under Soya Bean by size of holding 2005/2006 – 2011/12 Ethiopia

Year	Size of holding (Hectares)							
	<0.10	0.10-0.50	0.51-1.00	1.01-2.00	2.01-5.00	5.01-10.00	>10.00	Total
2005/06	*	*	274.00	895.00	1,902.00	207.00	-	3,327.00
2006/07	*	116.00	1,021.00	2,456.00	2,108.00	*	-	6,352.00
2007/08	*	252.00	745.00	3,185.00	2,498.00	*	*	7,807.00
2008/09	*	154.00	844.00	1,786.00	*	259.00	-	6,236.00
2010/11	*	600.00	1,399.00	5,248.00	3,412.00	*	-	11,261.00
2011/12	1.00	220.00	1,229.00	4,154.00	6,278.00	*	*	19,397.00

Source: Central Statistical Agency of Ethiopia

* These estimates could not be reported in this table because of high
– Data not available

Area covered by the crop from 2005/06 to 2011/12 production period shows uneven holding which moves up and down. This shows once the farmers produce the crop they ignore it coming year because of market problem of the crop. They even don't know how to locally process it. This leads them to carelessly engage in production of the crop. This needs attention of future intervention either by government or nongovernmental organizations on area of improving farmers' ability of processing soya bean, creation of favorable market condition for this crop.

Conclusion

Access to local markets appears to be the main constraint in many developing countries in the tropics and sub-tropics where local soybean production could improve farmer incomes and the sustainability of the production system. The problems of producing soybean is not only limited to market access but also low productivity and production, lack of processing facilities, lack of capital to increase production and no market information system for effective agricultural marketing. The high production record was obtained in 2012/13 production season, which is 636,531.01 quintals. But high yield (20 quintals per hectare) was obtained during 2013/14 production season. Area covered by the crop from 2005/06 to 2011/12 production period shows uneven holding which moves up and down. This shows once the farmers produce the crop they ignore it coming year because of market problem of the crop. They even don't know how to locally process it. This leads them to carelessly engage in production of the crop. This needs attention of future intervention either by government or nongovernmental organizations on area of improving farmers' ability of processing soya bean, creation of favorable market condition for this crop.

Acknowledgment

The author wishes to thank the reviewers of this article for their thorough and valuable feedback concerning what is discussed here above.

References

[1] Whigham,D.K. "Soybean Production, Protection, and Utilization", Proceedings of a Conf. for Scientists of Africa, The Middle East, and South Asia, University of Illinois International Soybean Program Urbana, Illinois 61801, **(1974),** pp.1-266

[2] Franklin W. Martin. "Soybean, Why Grow Soybeans", ECHO, 17391 Durrance Rd, North Ft. Myers FL 33917, USA, **(1998),** pp.1-5

[3] Ambitsi, N. E. Onyango and P. Oucho. "Assessment of Adoption of Soya Bean Processing and Utilization Technologies in Navakholo and Mumias Divisions of Western Kenya", Kenya Agricultural Research Institute, **(2007),** pp.1434-1438

8

[4] Dugje, I.Y., L.O. Omoigui, F. Ekeleme, R. Bandyopadhyay, P. Lava Kumar, and A.Y. Kamara. "Farmers' Guide to Soybean Production in Northern Nigeria", International Institute of Tropical Agriculture, Ibadan, Nigeria, **(2009),** pp.1-16

[5] Collombet R.N. "Investigating soybean market situation in Western Kenya: constraints and opportunities for smallholder producers", Wageningen University, **(2013),** pp.1-43

[6] Thoenes. P. "Soybean International Commodity Profile, Markets and Trade Division Food and Agriculture Organization of the United Nations", **(2014),** pp.1-25

[7] Centre for Development Innovation. "Policy Brief Developing new value-chains for soybean in Ethiopia", Wageningen UR, Netherlands, **(2012),** pp.1-10

[8] Bezabih Emana. "Market Assessment and Value Chain Analysis in Benishangul Gumuz Regional State, Ethiopia", SID-Consult-Support Integrated Development, **(2010),** pp.1-84

[9] Chianu J. "Soybean (Glycine max) promotion for improved nutrition and soil fertility in smallholder farms, East Africa", Tropical Soil Biology and Fertility Institute of the International Centre for Tropical Agriculture (CIAT), **(2006)**

[10] Ogema M.W, Ayiecho P.O, Okwirry J.J, Kibuthu I., Riungu T.C, Karanja D.D, Ng'ang'a C.N, Ocholla P. and Ireri EK. "Oilcrop Production in Kenya: Vegetable Oil/Protein System programme working paper series", Egerton University, Njoro, Kenya, **(1988)**

[11] Kasasa P, Mpepereki S. and Giller K.E. "Quantification of nitrogen fixed by promiscuous soybean varieties under Zimbabwean field conditions", in forum 4 working Document 1, Programme extended abstracts, Fourth regional meeting of the Forum for Agricultural Husbandry, Lilongwe, Malawi **(2000).**